Exploring
AI ENGINEERING
Careers

TITLES IN THIS SERIES INCLUDE

Exploring AI Engineering Careers

Exploring Biomedical Engineering Careers

Exploring Chemical Engineering Careers

Exploring Civil Engineering Careers

Exploring Mechanical Engineering Careers

EXPLORING ENGINEERING CAREERS

Exploring AI ENGINEERING Careers

Leanne Currie-McGhee

ReferencePoint Press

San Diego, CA

For more information, contact:
ReferencePoint Press, Inc.
PO Box 27779
San Diego, CA 92198
www.ReferencePointPress.com

LIBRARY OF CONGRESS CATALOGING-IN-PUBLICATION DATA

Names: Currie-McGhee, Leanne, 1971– author
Title: Exploring AI Engineering Careers / by Leanne Currie-McGhee.
Description: San Diego, CA : ReferencePoint Press, Inc, 2027. | Series: Exploring Engineering Careers | Includes bibliographical references and index.
Identifiers: LCCN 2025048158 (print) | LCCN 2025048159 (ebook) | ISBN 9781678212803 library binding | ISBN 9781678212810 ebook
Subjects: LCSH: Artificial intelligence--Engineering applications--Vocational guidance--Juvenile literature | Systems engineering--Computer programs--Vocational guidance--Juvenile literature
Classification: LCC TA347.A78 C87 2027 (print) | LCC TA347.A78 (ebook) | DDC 006.3/1--dc23/eng/20251222
LC record available at https://lccn.loc.gov/2025048158
LC ebook record available at https://lccn.loc.gov/2025048159

CONTENTS

AI Engineer at a Glance

Working Conditions

Mostly office or remote work

Certification and Licensing

Voluntary certifications in AI, machine learning, and/or cloud computing

Personal Qualities

- ☑ Creative
- ☑ Analytical
- ☑ Curious

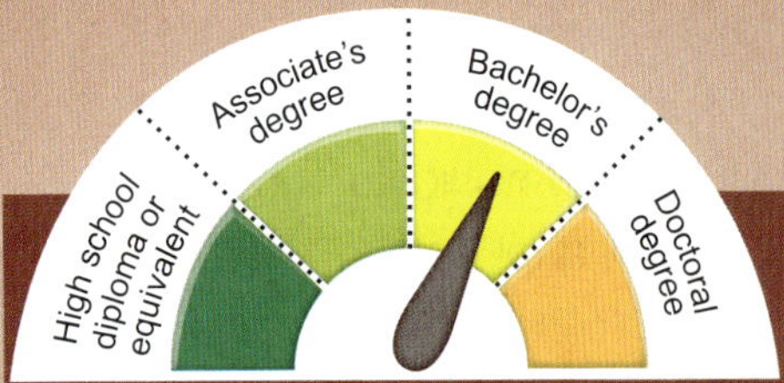

Minimum Educational Requirements

Bachelor's degree in computer science or related discipline

Pay

Average salary of $101,752 in July 2025

Future Job Outlook

Growth by 2027

40%

35,445

AI-related positions in the United States in 2025, according to Veritone

Number of jobs

Note: Please see page 63 for source information.

The New Frontier

Not long ago, artificial intelligence (AI) was considered science fiction. But today, AI is very much part of our everyday lives. From self-driving cars to lifesaving medical systems, AI is transforming nearly every industry. In the automotive industry, Tesla's AI-powered Autopilot system can change lanes on the highway without the driver touching the steering wheel. And in the medical field, IBM's Watson for Oncology uses algorithms to scan images faster than any human and assists doctors in diagnosing cancer. Behind these innovations are AI engineers. These are the people who design, build, and update the smart systems that can see, learn, analyze, and make decisions. Their work is driving one of the fastest-growing and most exciting career fields of today.

Experienced AI engineers have become so valuable that a few have become the focus of bidding wars between companies seeking top AI talent. Ruoming Pang, originally from Shanghai, China, earned his undergraduate degree at Shanghai Jiao Tong University, followed by a master's degree and doctorate in computer science from the University of Southern California and Princeton University, respectively. Early in his career, Pang worked at Google on its speech-recognition system before moving to Apple. At Apple, he led the team that developed the AI systems behind Apple Intelligence, which is the "brain" of the AI features built into Apple products. Pang's expertise made him one of the most sought-after AI engineers in the world.

After an intense bidding war in 2025, Pang left Apple to join Meta's AI research group, Meta Superintelligence Labs. His compensation package at Meta, according to the financial and technology news company Bloomberg, could exceed $200 million over multiple years. This leads to the following question: Why would a company pay so much for one engineer? The answer: Because in the race to build the most powerful AI systems, having the person best suited for the job can help a company move ahead of its competition.

> **"[The companies] want to move as fast as possible, so they are willing to pay a lot for candidates with specialized and complementary expertise."[1]**
>
> **—Ariel Herbert-Voss, CEO of cybersecurity start-up RunSybil and a former OpenAI researcher**

Pang is not the only AI professional receiving offers like this. Meta, OpenAI, and Google DeepMind have reportedly offered up to $20 million per year for experts in the AI field. Ariel Herbert-Voss, chief executive officer (CEO) of cybersecurity start-up RunSybil and a former OpenAI researcher, explains that the companies "want to move as fast as possible, so they are willing to pay a lot for candidates with specialized and complementary expertise."[1]

Why Are They Needed?

Although most AI engineers will probably not be earning millions annually, companies do pay well for AI experts. According to the online business news publication *Business Insider*, in 2025 an AI engineer at Meta made between $165,000 and $440,000. These levels of pay reflect AI's rising status as an essential technology. Because companies need AI to stay competitive, there are many opportunities for newly trained AI engineers to enter the field. They are needed to design algorithms, which are the step-by-step sets of instructions that guide how AI models analyze data and solve problems. They train AI models by feeding them data so the models learn to recognize patterns and improve performance. Additionally, these engineers build AI systems, which integrate models with interfaces and infrastructure to make data-driven decisions, often faster and more accurately than humans.

An AI model learns patterns from data to make predictions or decisions. A system is a completed setup of one or more models, along with user interfaces. The system is what people use to accomplish real-world tasks. For example, a customer service chatbot is a system that uses natural language processing, which is a type of AI that enables computers to understand and respond to human language. It also uses models to interpret what the questions mean and generate accurate responses to the questions. The system includes a user interface in the form of a chat window. Users type their messages in the window, and a backend system retrieves customer account information from company databases to answer those questions.

AI systems are used in health care, transportation, entertainment, finance, and even sports analytics to improve efficiency, make processes faster, or personalize the user experience. The systems can help doctors analyze medical images, allow self-driving cars to respond to road signs, or recommend songs based on a listener's habits. Apps such as Microsoft's Seeing AI can even describe text, objects, and people aloud to help individuals with vision impairments interact with the world. With every new advancement, the demand for skilled AI engineers continues to grow.

Because AI affects nearly every part of modern life, newcomers in the field will likely have many chances to gain a position with the potential to advance. New job postings for April 2025 increased 184 percent compared to the same month the previous year, according to the Computing Technology Industry Association, a trade association for the information technology (IT) industry. Jobs that did not exist twenty years ago, such as a machine learning expert, computer vision engineer, and natural language processing engineer, are among some of the hottest positions in technology.

Not Just the Money

While the money in AI engineering is certainly appealing, most professionals say that is not why they chose the field. For many, it is the chance to make a real difference in the industries they work in. This can include developing lifesaving medical technologies, designing climate models to predict and help prevent environmental problems, or creating accessibility tools to help people with hearing or vision impairments. AI engineers are tackling real-world problems that impact millions of people every day.

Alessandro Lamberti is a machine learning and computer vision engineer who works in Rome, Italy. However, he did not start out in AI. While studying telecommunications engineering in 2020, he realized he wanted to shift his career path. To make that change, he enrolled in classes offered through an online platform. This helped him develop skills in AI and other tech, eventually leading to internships and full-time jobs in the field.

In his current role, Lamberti uses computer vision (a type of AI that teaches computers to "see" and understand video footage) to analyze sports events. His work helps companies estimate the ages and genders of attendees, assess the actual number of people attending versus tickets sold, and determine where people are located throughout the games. This lets companies understand how people behave at the events, which helps management make decisions such as where staff should be at entrances based on crowd size. Age and gender data also help companies customize marketing services. Overall, Lamberti's work allows companies to better plan for sports events.

In Lamberti's view, where a person went to college or what degree that person obtained is less important to a potential employer than being able to show strong skills and deep knowledge. "Companies care only up to a certain point (if any) that you did some courses or degrees," Lamberti explains. "They care more that you can do the work, so focus on learning other supplemental skills, and learn how to get to the next stage of your career!"[2]

Lamberti's work is just one example of the possibilities of AI. For teens who are curious, creative, and interested in technology, a career in AI is worth considering. There are many different AI engineer roles across a wide variety of industries. Some engineers focus on designing algorithms, others on testing and improving AI systems, and still others work with machine learning, For anyone who wants to be on the forefront of innovation and has an inventive mind and the drive to work hard, there is a place in this rapidly growing field.

What Does an AI Engineer Do?

Kairan Quazi is not your average teenager. He earned a degree in software engineering at age fourteen. His senior thesis focused on AI and predicting human speech. Immediately after his graduation, in 2023, he began work at the space technology company SpaceX as an AI engineer. Quazi was drawn to AI because of his love for technology and his ability to push the boundaries of what it can do.

AI is a technology that allows computer systems to perform tasks that typically require human intelligence, such as thinking, learning, and even making decisions. Often, AI creates systems that can accomplish these tasks faster than humans. AI systems use computer programs that are called models, which are trained on large amounts of data to recognize patterns and make predictions from these patterns. Quazi worked on AI systems at SpaceX that helped make the company's satellite internet network faster and more efficient.

Building on the AI expertise he gained at SpaceX, in 2025 Quazi made a change. He took a job in the finance industry with a firm that handles US retail stock trades. Today, Quazi works in quantitative finance (also known as quant finance); this field combines math, coding, and AI to help businesses and individuals make financial decisions. In this role, AI engineers such Quazi create computer models that predict stock prices, identify smart investment opportunities, or measure how risky a trade might be. "Quant finance offers a pretty rare combination: the complexity and intellectual challenge that AI research also provides, but with a much faster pace," says Quazi. "I'll be able to see measurable impact in days, not months or years."[3]

The Work of an AI Engineer

AI engineers like Quazi use programming, data analysis, and specialized AI tools to design systems that help machines perform tasks, such as recognizing objects, understanding language, and predicting outcomes. They apply different techniques to create AI models and systems. One such example is machine learning. Machine learning is a method used by AI engineers to teach computers to learn from data and get better at tasks over time. Quazi's work in quant finance involves using machine learning to teach the computer to analyze financial data and predict stock trends.

> **"Quant finance offers a pretty rare combination: the complexity and intellectual challenge that AI research also provides, but with a much faster pace."[3]**
>
> —Kairan Quazi, teen AI prodigy

Another area that AI engineers may specialize in is computer vision, which is used to teach computers to "see" and understand images and videos. Self-driving cars use computer vision to recognize pedestrians, other vehicles, and road signs. Another area of AI involves large language models (LLMs). These are programs that can understand and generate human language. ChatGPT is an LLM. It can answer questions and interact with users in a way that sounds like a real person. Sometimes AI engineers focus on multiple areas at once. For example, when building a mobile app that can take a photo when a person smiles, AI engineers first use computer vision to help the app detect a face and then machine learning to make the app recognize a smile and take the photo.

AI systems are used in different ways. They can be narrow, meaning they focus on a single task, such as answering questions or filtering spam. They can also be generative, where they create new content, such as text, images, or music. In either case, it is the job of the engineer to develop systems that use data to make decisions, recognize objects, generate and understand language, or a combination of these skills.

Building the Brains Behind AI

AI engineers design, develop, build, test, and update AI systems. Their focus is to use AI models to develop practical software that can solve problems or make predictions. For example, they could develop fraud-detection programs for banks. This type of program would recognize and alert the bank about unusual spending patterns in real time. For health care settings, AI engineers might create systems that help doctors identify early signs of disease that appear in medical scans. No matter what industry they work in, the goal is to build intelligent tools that make decisions and improve the way businesses or organizations operate.

To do this, AI engineers design algorithms, which are step-by-step instructions that tell a computer how to analyze data and respond to it. They also train models, which means feeding large amounts of data into a computer program so it can learn patterns and make predictions. AI engineers sometimes start from scratch, but more often they modify and update existing models using frameworks like TensorFlow or PyTorch, which provide prebuilt tools that work with the models.

Once an AI system is working, engineers must continue testing and improving it. They monitor how accurately it performs and adjust it by entering new data to make it "smarter" over time. For example, Netflix AI engineers created and continually update its computer models to study what viewers watch, skip, or rate highly. The models learn viewers' preferences and suggest shows or movies that they are likely to enjoy. The engineers modify the system to keep recommendations accurate and personalized. Their work allows Netflix to personalize each customer's experience and keep millions of viewers engaged. In the past, Netflix has reported that 80 percent of what people watch is discovered via the company's recommendation system, making the AI work vital to viewership.

AI in Space

Although AI engineers may not end up in space, their work just might. In 2025, Chinese AI engineers implemented an LLM named Wukong AI, which was used aboard China's Tiangong space station during in-orbit missions. Wukong AI is a chatbot with expertise in navigation and tactical planning. AI engineers trained it by inputting large datasets of aerospace flight data. It was used to support astronauts during a six-and-a-half-hour space walk, providing them with quick answers to technical questions and mission guidance.

This is not the only example of AI in space. AI engineers in the United States developed Astrobee, a free-flying robot system that uses AI. Astrobee was deployed in 2019 to help astronauts perform tasks on the International Space Station. In 2016, German AI engineers designed CIMON, an AI buddy that provides emotional support and information to astronauts. CIMON was sent to the International Space Station in 2018.

Machines That See

Some AI engineers specialize in computer vision and write computer code that tells the system how to process each image, breaking it down into shapes, colors, and patterns. They program models that can recognize objects, patterns, and movements. They train these models by inputting into the program massive numbers of pictures or videos, allowing the program to gradually learn what different images represent. AI engineers then integrate the trained models into real-world systems by connecting them to devices or software that use the data in real time. As an example, they might link, via programming, a computer vision model to a camera system in a self-driving car. Doing this enables the computer to recognize traffic lights, pedestrians, and road signs. It then quickly analyzes what it sees and tells the car's control system whether to brake, accelerate, or slow down.

AI engineers are adapting models for systems that have all sorts of practical uses. For example, researchers at the University of Florida developed drones equipped with cameras to take detailed photos of hemp crops. In 2023, AI engineers who worked on this project created systems that analyzed the images from the drones and determined plant size, health, and growth. This allowed the system to point out areas where plants were lacking nutrients or receiving more than they needed. Farmers used the

AI engineers develop sophisticated software that can solve problems and make predictions. In health care settings, that could involve AI that helps doctors identify early signs of disease that show up in medical scans.

information needed to make knowledgeable decisions about fertilizing the crops. This technology helps crops grow healthier and saves time, money, and resources.

In another example, AI engineers in Madhya Pradesh, known as the "Tiger State of India" because it has the highest tiger population in the country, are currently using computer vision to develop a system that will protect wildlife. The system the AI engineers are developing "will capture images of tigers, leopards, and other big animals, analyze the visuals, and instantly send alerts to train drivers about animal presence near tracks,"[4] explains project leader Saurabh Das. The engineers design and train the computer models that make this process possible. They program the AI to recognize the shapes and movements of large animals in the images collected by sensor-based cameras powered by solar energy or batteries. Once trained, the AI program can rapidly analyze new images automatically and send alerts within seconds if animals are detected near tracks. By combining computer vision and machine learning, these engineers are helping reduce collisions between trains and wildlife while showing how AI can be used to protect the environment.

Large Language Models

Just as some AI engineers focus on helping computers recognize and make decisions based on what they "see," others focus on helping computer systems process language and respond. These engineers work with LLMs to make practical applications that read, write, and respond like humans. To do so, AI engineers train LLMs on huge collections of text, including books, websites, articles, and more, so the models can learn how people use language. Over time, the models recognize patterns in how words and ideas fit together. The engineers add coding to enhance the capabilities of the models. This training allows the model to learn to answer questions, write essays, summarize information, or even help write computer code.

Once the LLMs are trained, AI engineers integrate them into real-world systems. As an example, an AI engineer might use an LLM to design an airline customer service chatbot that will answer questions, track reservations, and provide travel updates. Engineers connect the chatbot using application programming interfaces (APIs), which are digital links to company databases, so the model can retrieve accurate information and respond to customers.

AI engineers are working with LLMs in several other industries, including education. In 2024, researchers at the University of California, San Diego, developed an AI tutor to support students in their courses. The engineers started with an existing LLM, one that is similar to ChatGPT, and customized it by inputting course materials, including class notes and podcasts. This taught the model the specific content, terms, and examples from the course. The engineers also programmed the model with

"Our goal is to create a tutor that is available any time and anywhere."[5]

—Mohan Paturi, professor in the Department of Computer Science and Engineering at the University of California, San Diego

prompts and instructions so the AI program could respond to students with step-by-step instructions, ask them questions, and even give encouragement. "Our goal is to create a tutor that is available any time and anywhere. Long term, we want to build [a system that] other instructors can customize to meet course-specific needs,"[5] says Mohan Paturi, one of the lead researchers on the project.

AI engineers like Paturi are helping to shape how the world uses technology. Whether these AI engineers are building safer self-driving vehicles or creating personalized AI tutors, they are using AI to impact daily life in positive ways. The possibilities for AI engineering careers are as varied as the projects already underway and the ones that have yet to be imagined. For teens who are curious, creative, and critical thinkers, becoming an AI engineer provides a chance to build the future.

CHAPTER TWO

How Do You Become an AI Engineer?

Jason Crandall knew he wanted to work at the leading edge of technology. After earning a degree in computer science in 2022, he completed an internship and then got a full-time job as a software development engineer. Around this time, he began experimenting with AI tools such as ChatGPT and quickly realized that AI was about to transform the IT industry. "I was not one of those doom and gloom folk preaching that the rise of AI is the fall of industry," says Crandall. "I was one of the people that saw this just as an evolution in tech, a step in a direction to change the way we work."[6]

Realizing he wanted to be part of that evolution, Crandall taught himself how to use existing AI tools, built a chatbot, and connected it to live data sources. After showing this and other projects to his employer, he was offered the opportunity to become the company's first AI engineer. Crandall accepted, and today he builds and deploys AI applications based on LLMs.

> "I was one of the people that saw this just as an evolution in tech, a step in a direction to change the way we work."[6]
>
> —Jason Crandall, AI engineer

Crandall combined his IT background and drive to move forward with actively learning new technology to become an AI engineer. Teens interested in this path can begin preparing early by taking science, technology, engineering, and mathematics (STEM) courses in high school; seeking out relevant internships in college; and committing to continually learning new skills as the technology evolves.

High School Preparation

For teens interested in AI, high school is the time to start building skills and knowledge. Students considering this career should take advanced math and science courses, such as algebra, calculus, statistics, and physics—all of which contribute to the basic knowledge needed for understanding algorithms and data analysis. Additionally, taking computer science and programming classes are key to getting a good start in this field. Many high schools offer courses in the coding languages Python and Java as well as advanced computer science courses, all of which are good choices. Even if schools do not offer these opportunities, there are online platforms that offer affordable and sometimes free ways to learn.

Teens can also prepare through extracurricular activities, such as joining a coding or robotics club or attending STEM career fairs, to learn about future options. Self-initiated projects, such as building a chatbot, also add to a teen's experience and knowledge.

Hackathons are another way to learn and get experience. Hackathons are timed events where people work in teams to quickly design and prototype solutions to specific technical challenges. Students from El Dorado High School in Texas gained knowledge and hands-on AI experience in April 2025 as participants in a hackathon held at the University of Texas, El Paso. El Dorado's team competed in the event to create an AI solution that addressed an education-related challenge. Using the programming language Python, the team designed an AI-powered program that analyzes budget spreadsheets and offers real-time analysis to help school leaders make financial decisions. Team member Damian Wood said he had no experience with Python until the hackathon: "This event honestly taught me how to use Python." Referring to his newfound knowledge and the experi-

> **"This [hackathon] event honestly taught me how to use Python."[7]**
>
> —Damian Wood, high school student

Teen Innovator

When Zach Yadegari started teaching himself the ins and outs of computer programming and AI at age seven, he did not expect it to be on a fast track to his own AI company. But that is exactly what happened. After teaching himself programming, he moved on to creating apps and building gaming websites. Somehow, he managed to balance his schoolwork with coding camps and hackathons to keep up with the latest technologies.

In 2024, while in high school, Yadegari developed Cal AI, an app that uses AI to estimate nutritional information from a simple photo of a meal. Launched in 2024, it became an overnight success. Within six months, Cal AI surpassed 1 million downloads, and by May 2025 it was generating $34 million a year in revenue. Toward the end of high school, Yadegari said success had not changed him. "I think the most grounding is that I still go to classes and still have to raise my hand to go to the bathroom," he said in the spring of 2025. "I'm still a normal high schooler. I'm going to prom in a couple of months."

Now the CEO of Cal AI and a student at the University of Miami, Yadegari attributes his achievements to self-motivation. By learning on his own and putting that knowledge into practice, he has already developed a thriving career.

Quoted in Jennifer McLogan, "18-Year-Old from Long Island Creates Cal AI Calorie Counting App Worth Millions," CBS News, May 2, 2025. www.cbsnews.com.

ence of presenting the project to the judges, Wood added, "It's really going to help me get a job later on."[7] Like many teens who take part in these kinds of activities, Wood came away with stronger technical skills and with experience in developing AI for real-world needs.

College Plans

Most AI engineers working in the field today obtained degrees in computer science, data science, statistics, mathematics, or software engineering. Many colleges and universities now offer specialized majors in AI or in courses aligned with AI, such as machine learning or robotics. In 2024, for instance, the University of Pennsylvania announced that its engineering department was offering a new bachelor of science degree in AI. "The timing of this new undergraduate program comes as AI poses one of the most promising yet challenging opportunities the

High school students who are interested in AI engineering careers will benefit from taking math, computer science, and programming classes. Self-initiated projects, such as building a chatbot, can also help develop skills.

world currently faces,"[8] says Vijay Kumar, the dean of the university's engineering department. Similarly, in the fall of 2025 the University of California, San Diego, launched a new AI major within its Department of Computer Science and Engineering. The program focuses on teaching students how all branches of AI, including machine learning, robotics, and computer vision work. The program also teaches the real-world impacts of AI on people and society.

No matter what the major, students who are leaning toward a career in AI should take courses in programming languages such as Python and in algorithms, data science, LLMs, and machine learning. Students who have an interest in a specific field—such as health care, finance, or automotive engineering—should also plan on taking classes that relate to those industries. Understanding the specific challenges within an industry is key to knowing how AI can be applied to create meaningful solutions.

College internships are a good way to get experience and make connections with people in the field. Students can look for opportunities through their university's career center, attend career fairs, and network with professors who may have connections in the industry. Many companies offer summer internships—some paid, some unpaid—for computer science or AI-related roles.

In addition to internships, students can gain experience with AI by contributing to open-source AI projects. Doing this offers more than hands-on experience. It also provides a way to demonstrate skills to potential employers. As an example, fast.ai, an open-source platform, provides a way for those new to AI to work with real-world applications and projects, such as creating a chatbot.

College research projects are another way to build skills and apply those skills toward exploring real-world problems. While a student at the Massachusetts Institute of Technology (MIT) in 2018, computer scientist Joy Buolamwini discovered that facial recognition software could not detect her face. Buolamwini is Black. "Face detection wasn't really detecting my face until I put on a white mask," she says. "It was Halloween time, I happened to have a white mask around. Pull on the white mask, the face of the white mask is detected. Take it off, my dark-skinned face, the human face, the actual face, not detected. And so this is when I said: hmmm what's going on here?"[9] This discovery led Buolamwini to study facial recognition systems more deeply, and she found that they often fail on darker-skinned women. She created more inclusive datasets to input into facial recognition models. IBM is one of the

The Newest Major

The growth of AI has been so dramatic that colleges and universities have created dedicated AI majors and programs. In 2018, Carnegie Mellon University in Pennsylvania became one of the first schools to offer a bachelor of science degree in AI. By 2025, at least fourteen US universities, including Purdue University, the University of Pennsylvania, and MIT, had created an AI major. During the 2025–2026 academic year, 328 students were enrolled in MIT's AI major.

This trend is also taking place in other countries. In 2019, the Mohamed bin Zayed University of Artificial Intelligence, a graduate-level, research-based university in the United Arab Emirates was established to focus entirely on AI. The university offers programs in areas such as machine learning, computer vision, natural language processing, and robotics. As of fall 2025, the university had enrolled over seven hundred students from more than forty-seven countries.

companies that used her work to improve the ability of its AI systems to recognize different skin tones. With more accurate and less biased algorithms, the AI facial recognition systems improved.

Certifications Enhance a Résumé

Though licensing is not required to become an AI engineer, certifications can strengthen an individual's résumé. Platforms such as Google, Microsoft, and IBM offer certificates in machine learning, AI tools, and cloud computing. Online learning platforms such as Coursera and Udacity provide programs that teach about the practical applications of AI. These certifications demonstrate to employers a person's initiative and emphasize specialized knowledge. For those switching from another field to a career in AI development and engineering, these courses provide a way to gain initial knowledge.

Certifications, degrees, and education can help aspiring AI engineers to demonstrate the necessary technical and mathematical knowledge and skills for the field, but breaking into AI engineering requires more. Employers want to see that job candidates also have the ability to apply knowledge and skills in the real world. This is why developing your own projects or collaborating with others is so important. Together, your education and experience will show you are ready to use your AI skills to solve problems and make an impact.

What Skills and Personal Qualities Matter Most—and Why?

When Sameer Maskey was growing up in Nepal, he spent hours working on his computer, despite the fact that he had to deal with spotty electricity in his neighborhood. His curiosity about what technology could do eventually led him to pursue higher education in the United States. In 2005, he earned a bachelor's degree in mathematics and physics from Bates College in Maine. "The first time I heard about [AI]," Maskey recalls, "was in my first year undergrad when I was thinking of building talking machines. That was my dream, building a machine that can sort of converse with you."[10]

> "That was my dream, building a machine that can sort of converse with you."[10]
>
> —Sameer Maskey, founder of Fusemachines

Maskey's vision for what AI might achieve motivated him to continue his studies. He eventually earned a doctorate in computer science, with a specialty in AI and machine learning, from Columbia University. In 2013, he founded Fusemachines, a company focused on using AI to expand access to education and opportunities in underserved regions worldwide. The company offers full scholarships, AI certifications, and assistance with job placement to individuals in countries such as Nepal and Rwanda.

Maskey developed the technical skills he needed through coursework and independent projects, but it was his drive and vision that helped him achieve such success.

Aspiring AI engineers must master a combination of technical skills, including data science, natural language processing, data handling, and machine learning, and soft skills such as creativity, vision, and initiative. Together, these abilities develop AI professionals who are capable of designing cutting edge technology in responsible and meaningful ways.

Technical Skills Every AI Engineer Needs

From a technical perspective, AI engineers must have solid programming skills. Languages such as Python, Java, and C++ are essential for working on AI systems. They allow engineers to translate abstract ideas into models by coding the algorithms. Also important is the ability to work with data. AI models use massive datasets, so AI engineers need to understand how to collect, clean, and manage data using tools such as Structured Query Language (SQL). Familiarity with data frameworks that process and analyze large datasets is also needed.

Because mathematics and statistics are the basis of AI, coming into the field with proficiency in linear algebra, calculus, probability, and statistics is critical for building, updating, and evaluating machine learning models. A strong grasp of machine learning fundamentals is necessary because it provides the basis for how algorithms learn from data to form models. AI engineers need to be proficient with popular frameworks and libraries such as TensorFlow, PyTorch, and Keras. This will enable them to use ready-made models as well as design and build custom model architecture. This combination of strong mathematical knowledge, machine learning understanding, and practical framework skills gives AI engineers the skills they need to develop accurate and efficient AI systems.

AI engineers also need skills that allow them to manage and evaluate information used in AI models. For example, they must understand retrieval techniques, such as vector search, that help LLMs rapidly find needed data. They also need to know how to evaluate models for accuracy and response times. These tasks are done with AI monitoring tools that track performance, detect errors, and visualize results. These evaluations are necessary because AI tutors and virtual assistants can provide inaccurate or misleading information. The monitoring tools ensure AI engineers are able to evaluate system performance and update as needed. Finally, because AI apps are often used by millions of people at once, they require large computer power and data storage. To deal with this, engineers need to know how to run and scale the models in the cloud. The cloud is a network of remote internet servers that allows companies to store, manage, and process massive amounts of data. AI engineers need to have a working knowledge of cloud platforms and the tools that enable different computer programs to work together.

People Skills Also Matter

While technical skills are essential for anyone who wants to work as an AI engineer, people skills are also important. Empathy, emotional intelligence, ethical judgment, creativity, and leadership are uniquely human qualities. According to a 2025 MIT study, these qualities are not easily transferred to AI. The researchers emphasized that productivity and innovation can only result when these qualities are paired with technical skills. As Fei-Fei Li, an AI pioneer and professor says, "There's nothing artificial about AI. It's inspired by people, it's created by people, and—most importantly—it impacts people."[11]

Empathy and ethical judgment, for example, are vital when designing AI that directly affects people's lives. When developing a model for a hospital to manage patient care, engineers must consider the human impact of different decisions. Should the AI prioritize speed of diagnosis or thoroughness? Or, when building a system to help a company identify job candidates, how can the algorithm be designed to reduce bias? AI engineers who thoughtfully consider the human impact of their work are more likely to create products that actually provide a benefit to society.

A Visionary

The work of AI pioneer Fei-Fei Li shows how vision can change an entire field. Back in 2006, while an assistant professor at Princeton, Li had a bold idea. She wanted to use AI to give computers a human ability, a way to see the world. Li began to build ImageNet, a massive online database of millions of labeled photos. She used this to teach AI programs how to recognize everyday images like cats, cars, and trees. Back then, this was a groundbreaking accomplishment and pushed AI forward. Li continued to let her vision guide her career. She foresaw how powerful AI would become and realized it was important to find ways it could benefit people. Now, as codirector of Stanford's Institute for Human-Centered Artificial Intelligence, she focuses on using AI to improve people's lives in areas like health care, education, and the environment. Her long-term vision has pushed the field forward while ensuring people are served by AI's possibilities.

Creativity and vision are also crucial in this career. AI engineers need both as they design innovative solutions to complex problems and anticipate future needs and possibilities. For example, AI engineers at a production company that worked with Marvel Studios paired creativity with generative AI to produce watercolor-style opening credits for the Disney television series *Secret Invasion*. As for vision, Cinelytic, a company that provides analysis using AI, partnered with Warner Brothers in early 2020 to predict how the company's potential movies would do when released. The AI engineers developed models that used script data, historical trends of movie audiences, and other related data to determine box office potential. This insight helped company executives in deciding whether or not to approve production of new films.

Additionally, AI engineers must be able to communicate clearly with people who do not have a technical background. Clients usually know what they want done, but they do not necessarily have a background in the technological details that can make their project happen. The ability to communicate complex ideas more simply is important when dealing with clients, but this quality can also be a plus during a job interview.

Continuous Learning

The rapid rise of AI and the steady pace of advances in this technology make ongoing learning essential for any AI engineer. "AI is evolving at lightning speed. That might sound intimidating, but it's actually an opportunity to make continuous learning your superpower,"[12] advises AI consultant Nadeen Matthews Blair.

> "AI is evolving at lightning speed. That might sound intimidating, but it's actually an opportunity to make continuous learning your superpower."[12]
>
> —Nadeen Matthews Blair, AI consultant

Manoj Tumu realized early on the importance of continually adding new knowledge to what he had already learned. After earning both a bachelor's and master's degree in computer science, he got a job as a machine learning engineer at Amazon Web Services, where he gained hands-on experience with AI projects. Beyond his job, Tumu actively pursued additional learning opportunities. He stayed updated with the latest advancements through his own study and used these newly learned techniques in his work. In 2025, twenty-three-year-old Tumu landed a job as a machine learning engineer on Meta's advertising research team. In this position, his job is to develop AI that understands and predicts

Demonstrating Ethics in AI

In the AI field, ensuring professionals focus on ethics when they design technology has become so important that many companies have created ethical AI teams to guide their work. By focusing on ethics, these teams ensure that AI helps people and does not harm them. To do this, engineers have to think ahead and consider how their technology could be misused or accidentally hurt someone. Being ethical as an AI engineer also means building systems that are fair and honest and that protect people's privacy. Another big responsibility is preventing bias or discrimination from showing up in the data used to train AI. For example, AI engineer Timnit Gebru, who once co-led Google's ethical AI team, warned that LLMs can reflect or even increase bias that already exists in data used for training. This could result in an LLM producing stereotyped or unfair responses, such as linking certain jobs to a particular gender or race. Her research pushed major technology companies to consider how their AI models are trained in order to make them more fair. Today, being ethical is important to make sure AI does not cause harm to others.

customer needs so that the company can more accurately target their promotions. He carries out this work with the aid of cutting-edge AI models. Manoj's dedication to learning and skill growth has been a major part in advancing his career, leading to challenging, impactful roles in AI engineering.

Tumu's journey shows that success in AI is not the result of one specific degree or one job. It is a combination of skills, drive, and focus on continually expanding knowledge. These are the elements that can help lead to a career as an AI engineer.

What Is It like to Work as an AI Engineer?

The work of an AI engineer depends on the specific job, such as whether it focuses on LLMs or computer vision, but for most, it includes solving real-world challenges. For AI engineers working on machine learning projects focused on self-driving vehicles, one challenge is reducing vehicle accidents. These engineers often work on teams that teach self-driving cars how to avoid collisions.

This is accomplished through a process that involves training models on huge datasets collected from vehicles right before past crashes. Part of the job focuses on inputting the collected data into the model. The engineer team inputs information into a system that stores and organizes huge amounts of data. Another system processes and analyzes the datasets so they can be used to train models and run simulations. The data must be prepared and tested carefully so the AI system can learn to make smart, safe decisions.

Self-driving cars operate through a system of sensors that act like the car's eyes and ears. These sensors, such as cameras and radar, collect data about the vehicle's surroundings, including nearby cars, lane markings, and pedestrians. The AI system then processes this information to make driving decisions. If a sensor malfunctions or sends poor-quality data to the model, the AI system could misinterpret the situation and make an incorrect decision, such as braking too late or steering in the wrong direction. Engineers who work on these systems constantly test and calibrate the sensors to ensure they are providing accurate and reliable input. They also evaluate how the AI system interprets this data, ensuring it can recognize when a sensor is providing faulty information. This careful testing helps prevent accidents and keeps autonomous cars operating safely, even when part of the system fails.

Another part of the work for these AI engineers involves simulations. Because testing crashes on real cars is not practical, they run computer-based crash simulations to observe how well the AI model reacts. Sometimes their job is to help the model learn to make survivability decisions, which basically means they teach the AI how to make the safest possible choice in dangerous crash situations. For example, the system might have to decide between swerving into a telephone pole or hitting a parked car. At increased speeds, the outcome of such a crash can change dramatically, so the engineers feed the AI model crash data at different speeds and angles, then monitor how the system responds. If the AI model makes a poor choice, the engineers improve it by adding more data from similar crashes, which helps the model adjust its rules and make better decisions in the future. Over time, it becomes better at consistently making safer decisions. Data and model testing is how engineers focused on vehicle collision spend the day. However, other AI engineers working on self-driving vehicles may focus on natural language tools, robotics, or health care, so their specific tasks vary. For all, the goal is to design AI solutions that provide safe, effective, and innovative results.

A Day in the Life

Kriti Goyal, twenty-eight, is a machine learning engineer in Seattle. She spends her workdays developing real-world solutions with AI. Having grown up in India, she immigrated to the United States and, in 2022, earned a master's degree in computer science from the University of Wisconsin–Madison. There, she conducted research in natural language processing and computer vision. Today, she is part of an engineering, enterprise, and AI solutions team. The team builds the brains of AI systems so the models in the systems can learn from new data and make smart decisions. "I work on building the foundation of machine learning models, which means I build code that trains software to recognize unseen data and create patterns,"[13] explains Goyal.

Goyal structures her workday into three main segments that include research, collaboration, and hands-on coding. She starts her day by studying and analyzing new methods and technologies to enhance AI systems. The next part of her day involves meeting with team members and clients to ensure teams have what they need and understand priorities and to check that the solutions they are providing meet clients' needs.

> **"I'm lucky enough to spend most of my time being an individual contributor and focusing on coding."[14]**
>
> —Kriti Goyal, machine learning engineer

Finally, Goyal dedicates time to hands-on coding, actively developing and refining the AI models. This is the best part of her day. As she explains, "Everyone's favorite part is the third, which is basically hands-on building and coding. I'm lucky enough to spend most of my time being an individual contributor and focusing on coding."[14] Her work involves writing code that trains AI software to identify patterns in data, which is how models learn. Goyal's coding and analyzing models, research into new methods, and teamwork help bring new technology to life.

Work Environment and Pay

Although every AI engineer's day may be a little different, the settings they work in, the pay they can expect, and the pressures they deal with are similar no matter their field. Most AI engineers work full-time and either on-site or in hybrid situations, where they split their work between coming into the office on some days and working from home or elsewhere on others. Many industries have embraced hybrid models, especially after the pandemic reshaped work culture. However, some positions, such as research-focused positions, especially at universities or labs, may require more time in person with the computers and materials on-site.

> **"I think the major harm that comes is there's no time to think critically."[15]**
>
> —Ayodele Odubela, a data scientist and AI policy adviser

The pace of AI engineering work varies. A forty-hour workweek is typical, but longer hours are common as project deadlines approach or change. For example, an Amazon AI engineer, interviewed by the news network CNBC in 2024, described finishing his work for the week and looking forward to spending the weekend with visiting friends. Then he received a message about a project deadline being moved up to early Monday morning, which meant he missed out on time with his visiting friends.

In an industry where companies are racing to outperform each other, high-pressure situations like this are not uncommon. Another worker in AI, Ayodele Odubela, a data scientist and AI policy adviser, comments that the pressure and rush to get AI products completed does not always allow her time to consider the best options and produce her best work. "I think the major harm that comes is there's no time to think critically,"[15] Odubela says.

Dealing with these high-pressure jobs generally results in high pay. AI engineering is known for its high salaries, even at the entry level. New AI engineers often earn between $100,000 and $140,000 per year, according to a 2025 report by job recruiting

Protecting Kids Online

A global movement to make the internet safer for children is creating a growing number of jobs focused on AI safety. This, in turn, is opening opportunities in new areas for AI engineers. For example, companies are now building tools such as AI-powered filters that will work to get explicit content off kids' devices and use models to prevent young people from seeing age-inappropriate material. Companies are hiring child-safety analysts or AI trust and safety engineers to accomplish these tasks. Their job is to build and test systems that will flag harmful content, detect online predators, and attempt to stop online abuse before it happens. Roblox, a gaming company used by many young people, created a chat moderation system to protect youth. The system is powered by AI to recognize grooming, which is when an adult tries to gain a minor's trust and then manipulate them, and to alert parents. In these roles, AI professionals can use their technical skills to build tools that reduce potentially negative impacts on youth.

firm Mason Alexander. According to the same report, just three to five years of experience can earn an AI engineer $150,000 to $200,000, and senior engineers, especially those at major technology companies, reach average salaries of $200,000 to $300,000 or more. Total compensation exceeds those numbers when including bonuses and stock options.

Where Can AI Engineers Work?

AI engineers have the opportunity to find work in many different industries, at companies of all sizes, and in areas across the country and the world. In technology and software, major corporations like Apple, Google, and Microsoft lead AI innovation in areas such as Silicon Valley, Seattle, and Austin. North Carolina's Research Triangle has become a hub for health care and life sciences businesses that rely on AI. Finance and insurance firms

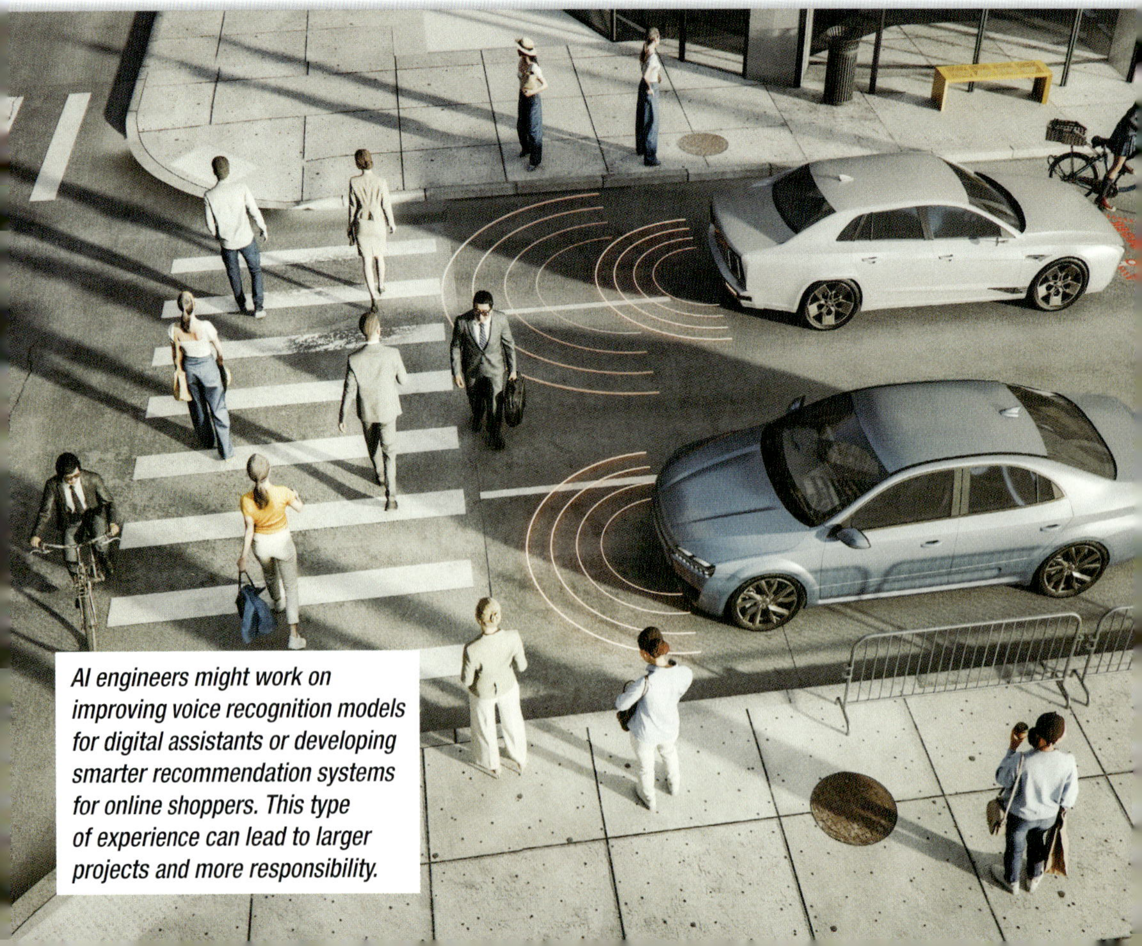

AI engineers might work on improving voice recognition models for digital assistants or developing smarter recommendation systems for online shoppers. This type of experience can lead to larger projects and more responsibility.

based in major cities such as New York and Chicago use AI to help find fraud in transactions and analyze investment risks. Even media and entertainment companies have begun to use AI as a part of content creation and enhancement of personalized experiences for audiences.

The companies that hire AI engineers range from small startups to global corporations. Specific roles are increasing in machine learning, systems engineering, and other technical areas as well as in complementary areas such as AI strategy, compliance, and ethics. Most experts foresee that qualified AI engineers will have little difficulty finding a place in the fast-changing AI job market.

Advancement and Other Job Opportunities

As in any profession, most AI engineers start their careers in entry-level roles. But that does not mean they have to stay at that level. A solid background in AI can lead in all different directions. It can lead to research and teaching. It can lead to new product development or other jobs in private industry. It can also lead to entrepreneurial ventures.

Whichever direction an AI engineer chooses, advancing in the field will require keeping up with the flood of new developments in AI technology. Orla Daly, chief information officer at the online training and digital learning company Skillsoft, believes that aspiring leaders in all technology fields need to continually keep up with AI developments. "In the past, as you progressed in your career in IT, you were less involved in the technology, and your role became more about leadership and management," she comments. "And while those leadership and management pieces continue to be super-relevant and important, you will also need to stay closer to technology . . . more than previously."[16]

One way to advance in AI engineering is to obtain a higher degree. Some AI engineers pursue master's degrees or doctorates in AI engineering or related disciplines. Others focus on specialized areas of study, such as natural language processing, generative AI, or computer vision. Getting a degree in another field, such as law, health care, or business, can also open doors to new opportunities. AI is being used more and more in all of these fields. A law degree combined with AI skills and experience, for instance, could open doors to an AI-related career in the legal field.

> "In the past, as you progressed in your career in IT, you were less involved in the technology, and your role became more about leadership and management. And while those leadership and management pieces continue to be super-relevant and important, you will also need to stay closer to technology."[16]
>
> —Orla Daly, chief information officer at Skillsoft

Research and teaching offer another option. Anima Anandkumar is a prominent AI researcher who has forged a career that includes teaching and research as well as working in the tech industry. She has degrees in electrical and computer engineering. That background led to teaching and research at the university level dealing with large-scale machine learning and mathematical models that help computers learn from massive amounts of data. Eventually, Anandkumar joined Amazon Web Services, where she helped develop machine learning tools that make it easier for companies to build and deploy AI systems. She returned to academia in 2018, becoming a professor of computing and mathematical sciences at the California Institute of Technology (Caltech). At Caltech, she leads the AI + Science Lab, focusing on AI algorithms that solve complex problems in physics, climate, and materials science.

She has also continued to work in private industry. As director of machine learning research at NVIDIA, one of the world's leading AI technology companies, she has created a joint project between Caltech and NVIDIA. This project involves an advanced AI-based weather model that can predict extreme weather events faster and more accurately than traditional methods.

Pathways in Industries

AI engineers who work in private industry can also find opportunities for career growth and advancement. Entry-level AI engineers at companies like Google or Amazon might work on improving voice recognition models for digital assistants or developing smarter recommendation systems for online shoppers. As they gain experience, these engineers can take on larger projects and more responsibility, sometimes becoming senior engineers or technical leads. At this level, they may design entire AI systems, mentor junior engineers, and collaborate with product and research teams to solve complex problems. For example, senior AI engineers at OpenAI collaborate across teams to improve LLMs like ChatGPT, finding ways to make AI responses safer, more accurate, and more useful. At this level, AI engineers not only build AI models but also think about their scalability (the ability to grow), efficiency, and reliability to ensure the technology works for millions of users.

Many AI engineers also choose to specialize in particular areas, such as generative AI or AI infrastructure. Some focus on AI ethics, which aims to ensure that AI systems are fair, transparent, and safe. Specialists in these areas are in high demand and often move quickly into senior technical positions.

With experience, some AI engineers advance to higher roles, such as staff engineer, principal engineer, or AI architect. In these positions, they are recognized as experts in their field. They might lead large teams, design complex AI architectures, or create new algorithms that improve accuracy and performance. Further advancement in industry includes leadership roles like engineering manager, director of machine learning,

Salary Growth

AI engineering not only provides opportunities to work on exciting projects but also offers impressive earning potential. According to the online learning site Coursera, entry-level AI engineers typically earn around $103,000 per year, and senior engineers and technical leads can earn between $157,000 and $178,000 annually. Gaining expertise in machine learning, natural language processing, or computer vision can also increase earning potential. For example, according to the job search site Glassdoor, in April 2025 the average salary for a machine learning engineer in the United States was $157,810. Overall, AI engineering offers strong salaries and benefits across the board. As engineers gain experience, specialize, and take on leadership roles, their earning potential grows significantly, reflecting the high demand for skilled professionals in this fast-moving field.

or chief AI officer. Individuals in all of these roles manage AI programs and make key decisions about how to use AI technologies to benefit the business. In these positions, technical knowledge is still important, but leadership, communication, and vision are also necessary.

Vijaye Raji began his career as a software engineer at Microsoft. Over time he has advanced to a senior engineer position at Facebook and, most recently, as the chief technology officer of applications at OpenAI. In this latest position, he heads product engineering for ChatGPT and Codex, which is an AI model that translates natural language instructions into computer code. Raji's responsibilities include overseeing the main systems and programs that make these AI tools work. "Our models are opening entirely new ways to build, and [Raji's] leadership will help turn that progress into safe applications that empower people with many new tools to improve their lives, help companies increase their impact and allow developers to build faster and better products,"[17] explained Fidji Simo, CEO of applications at OpenAI, when discussing Raji's hiring in September 2025.

Entrepreneurship Potential

AI engineering can also lead to careers outside of research or traditional industry roles, providing a pathway to entrepreneurship. AI engineers can apply the skills they develop in coding, machine learning, and problem-solving to building their own companies or launching innovative AI products. This path combines technical knowledge with creativity, business sense, and leadership. Anyone who wants to go this route would benefit from learning the basic principles of finance, accounting, and marketing.

The journey to one's own business in AI usually starts with a strong foundation in AI, such as designing algorithms, building models, and understanding how AI can solve real-world problems. Once AI engineers gain experience, they may spot openings in the market or problems that existing products do not solve. Some AI engineers choose to create start-ups, either alone or with a team of fellow engineers and entrepreneurs. As an example, Dmytro Lider earned degrees in computer science and software engineering, then began his career as a software engineer and project lead who specialized in language technology. In 2009, he cofounded Grammarly, an AI-powered writing assistance company that uses natural language processing to help users write more clearly and effectively. "The grand vision is to help people connect and understand each other, using artificial intelligence," Lider said of Grammarly. "We've always had the idea of helping people become overall better communicators."[18] Since it started, the platform has grown to serve millions of users worldwide.

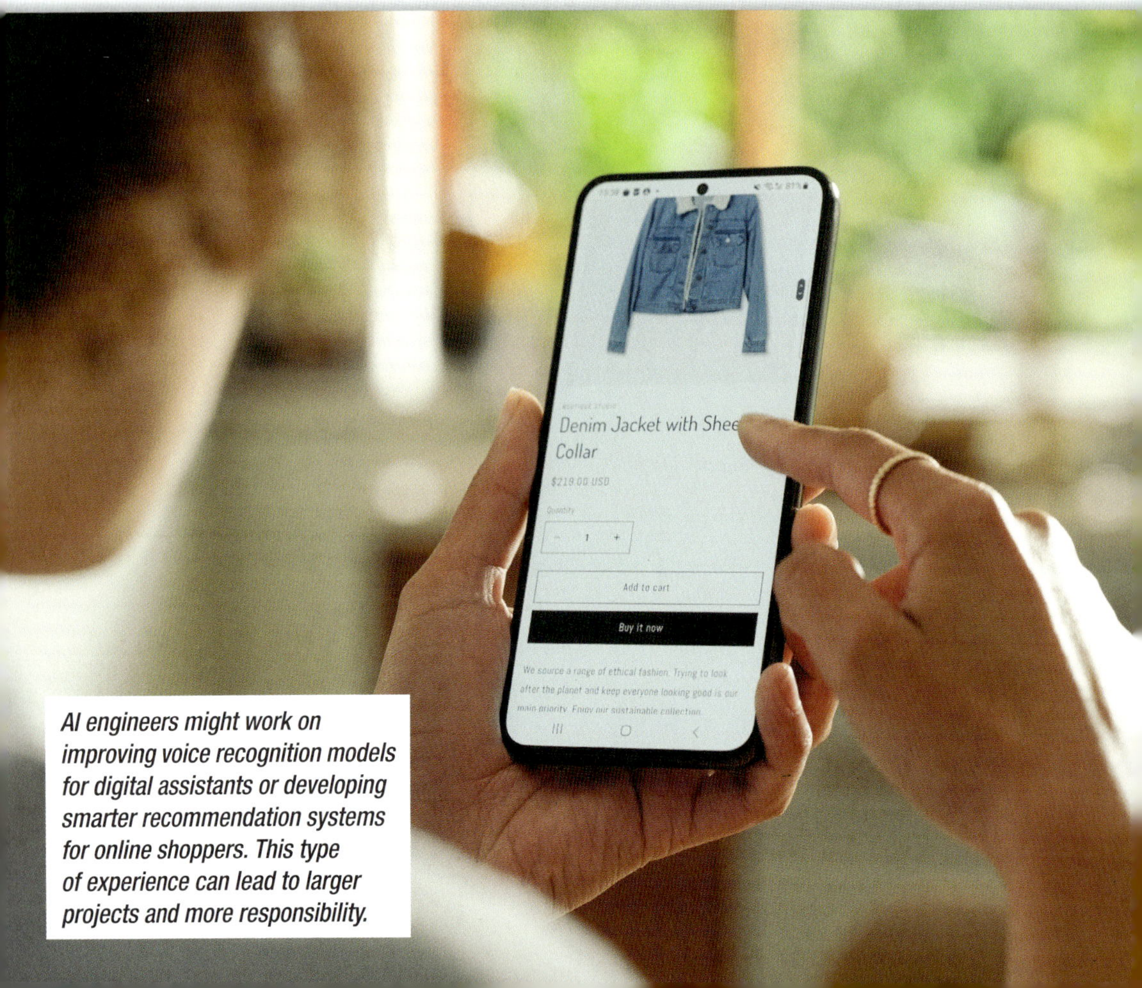

AI engineers might work on improving voice recognition models for digital assistants or developing smarter recommendation systems for online shoppers. This type of experience can lead to larger projects and more responsibility.

> "The grand vision is to help people connect and understand each other, using artificial intelligence."[18]
>
> —Dmytro Lider, cofounder of Grammarly

The possibilities for an AI engineer have expanded nearly as much as the technology itself. Whether following a path as a researcher, becoming an industry innovator, or building something entirely new, AI engineers can shape how people work, learn, and communicate. The key to success in all paths for AI engineers is to stay current with the evolving technology while maintaining a mindset that includes critical thinking and curiosity. Those who do might one day achieve the breakthroughs and accomplishments of industry leaders like Anima Anandkumar, Vijaye Raji, and Dmytro Lider.

CHAPTER SIX

What Does the Future Hold for AI Engineers?

The future looks wide open for AI engineers, as job opportunities for people with this expertise grow each day. While some people are concerned that AI will replace human workers in many industries, the need for professionals in the AI field itself is enormous. According to GrandView Research, a market research company, the global AI industry was valued at $279 billion in 2024 and is projected to reach $1.8 trillion by 2030. "We are living through a seismic shift, a moment as transformative as the Industrial Revolution of the 18th century or the advent of the internet at the dawn of the 21st," notes Muhammad Tuhin, a writer for *Science News Today*. "Artificial Intelligence (AI) is no longer just a sci-fi fantasy or a niche field of research tucked away in academia. It has erupted into the mainstream, permeating nearly every sector of the economy and society."[19]

> **"Artificial Intelligence (AI) is no longer just a sci-fi fantasy or a niche field of research tucked away in academia. It has erupted into the mainstream, permeating nearly every sector of the economy and society."[19]**
>
> —Muhammad Tuhin, a writer for *Science News Today*

Microsoft, Google, Apple, and other major technology companies are making billion-dollar investments in AI, and most other industries are searching for AI experts. According to a University of Maryland report, since 2022, job postings requiring AI skills have increased by 68 percent, even while other technology postings, such as those requiring coding skills, have decreased. Despite this rising demand, the technology training company Degreed found that only 3 percent of workers currently have expertise in AI. This gap means youth who start preparing for careers in AI now will have a big advantage over others.

The opportunities extend across industries. From health care to entertainment, companies need AI engineers to build models and systems that can improve businesses, enhance products, interact with customers, and adapt over time. Advancements like these are happening fast, giving AI engineers a wide variety of career options.

Health Care Sees Fast Changes

One of the industries experiencing the fastest AI growth is health care. AI is already making a difference in the industry today, with major potential for the future. At Addenbrooke's Hospital in Cambridge, England, Raj Jena, an oncologist and AI researcher, and his team built Osairis, an AI tool created using Microsoft technology. Normally, when doctors plan radiation treatment for cancer, it takes hours as they carefully map exactly where the beams should hit the tumor while avoiding healthy organs. Osairis, trained on thousands of past scans, can do this mapping in just minutes. As Jena explained in 2025, "We've gone quite quickly, from these approaches being just research to actually being plumbed into the clinic and helping patients get started on potentially life-saving radiotherapy more quickly."[20]

> **"AI holds enormous potential to transform care that could lead to substantial improvements in health."[21]**
>
> —Libor Pastika, a clinical research training fellow at Imperial College London

AI is also being developed to catch illnesses early. In England, engineers and researchers have developed a tool that analyzes electrocardiogram heart scans and predicts the risk of a patient developing type 2 diabetes up to ten years in advance. This tool was scheduled for clinical trials in 2025. Many in the medical field see this as just the beginning. According to Libor Pastika, a clinical research training fellow at Imperial College London, "AI holds enormous potential to transform care that could lead to substantial improvements in health."[21] This type of early warning system could give people time to change their lifestyles and potentially avoid serious illness later.

As AI takes on more complicated tasks in health care, such as planning treatments and predicting diseases, AI engineers are the ones who will build and update these tools. They will need to work with doctors and other medical professionals to design systems that help hospitals run more efficiently and find new ways to keep people healthy. From preventive detection to early diagnosis to managing treatments, AI engineers have the potential to transform health care.

Fighting Off Hackers

Another fast-growing area for AI engineers is cybersecurity. Every day, hackers try to break into utilities, hospitals, banks, and even school systems, and AI is becoming a powerful tool to fight back. Instead of waiting for an attack to happen, AI systems can monitor networks with millions of signals at once and instantly detect unusual logins, strange internet protocol addresses, or suspicious behavior before a system is hacked. Adding to this is that hackers are also starting to use AI, which means the need for skilled AI engineers in cybersecurity will only increase. According to KBV Research, the AI cybersecurity market is expected to reach more than $42 billion by 2028.

AI is also helping to stop phishing attacks, which are emails or messages designed to trick people into giving away passwords or personal information. Engineers are developing natural language processing systems that scan messages for suspicious wording, odd grammar, or urgent phrases like "act now" or "your account will be closed." Current AI tools can even check links against databases of known malicious sites and block them before users can click the links. In November 2022, Google reported blocking more than 231 billion spam and phishing emails in just two weeks. For AI engineers of the future, building smarter defenses like these will be essential to protect everyday users from scams.

Entertainment Gets Smarter

AI is even becoming a force in the entertainment industry. The use of AI in movies, music, gaming, and social media has dramatically increased in recent years. According to Market.US, the AI entertainment market in 2023 was valued at $17 billion. By 2033, that number is expected to reach $195 billion.

Streaming platforms such as Netflix and Spotify already use AI engineers to create models that will have their systems recommend shows or songs based on a user's viewing and listening habits. Filmmakers are also turning to AI tools to generate realistic special effects and speed up the editing process. Additionally, video game companies are using AI to design more lifelike characters that respond in unique ways to a user's actions, which gets users more immersed in the games.

Looking ahead, AI could even result in new forms of entertainment. Virtual reality developers are delving into AI to help create experiences where storylines immediately adapt to a player's choices. In sports entertainment, AI tools may eventually be used to generate highlight reels customized for a fan's favorite team or athlete. For AI engineers, options in the entertainment industry make use of artistic and technology talents.

AI Fighting Climate Change

AI is becoming a powerful way to work toward solutions for climate change. AI engineers are building systems that are able to analyze massive amounts of environmental data, from satellite images to ocean temperatures. This type of analysis can help scientists predict wildfires, track deforestation, and reduce energy use. Data centers, which store and manage computer servers, are huge energy users. The amount of electricity used to keep data center machines running and cool is often enough to power tens of thousands of homes. Google's DeepMind AI lab has developed an AI model that can reduce cooling system energy use in data centers by up to 40 percent. This change would produce fewer of the greenhouse gas emissions that contribute to climate change. This AI system accomplishes the reduction of energy use by analyzing data from thousands of sensors within the facilities, such as temperature and power, to predict future energy needs and change settings to optimize energy reduction. That means a career in AI could also mean a career in protecting the planet.

Even in Education

AI is also making its way into education, especially through engineers designing tools that personalize lessons for students. Adaptive platforms such as DreamBox and Coursera use machine learning to adjust lessons in real time. If a student struggles with a concept, the platform provides more practice, and if the student learns quickly, it moves them forward quicker.

Personalized tutoring is another area in which AI is being used. In China, the company Squirrel AI has created an AI system that identifies a student's weaknesses and then builds practice sessions instantly. In December 2024, Squirrel AI set a Guinness World Record for the most users to take an online mathematics lesson over a twenty-four-hour period. During the event, the sys-

Cybersecurity is a fast-growing area for AI engineers. AI systems can monitor networks and instantly detect unusual logins, strange internet protocol addresses, or suspicious behavior before a system is hacked.

tem generated 111,704 unique math exercises tailored in real time to each of the 112,718 participating students. "We use an AI algorithm to imitate the best teacher in the world," Squirrel AI founder Derek Haoyang Li explained. "Like Da Vinci plus Einstein together, to give every student in this world equal education, to have personal tutoring."[22] In the future, classrooms may pair human teachers with AI assistants, giving students 24/7 access to help while allowing teachers to focus on creativity and critical thinking. AI engineers could play a central role in shaping how education works for generations to come.

Challenges to Consider

While the future of AI engineering is exciting, experts also warn about challenges that engineers will need to solve. One of the biggest issues is bias, which happens when an AI system makes errors because the data used for training was incomplete or one-sided. For instance, a 2024 study by the University of Washington (UW) found that some AI hiring tools scored résumés differently just because of a person's name. If the AI linked a name to a particular race or gender, it sometimes ranked that person lower, even if they had the same qualifications as others. "Small companies could attempt to use these systems to make their hiring processes more efficient, for example, but it comes with great risks," says UW assistant professor Aylin Caliskan. "The public needs to understand that these systems are biased. And beyond allocative harms, such as hiring discrimination and disparities, this bias significantly shapes our perceptions of race and gender and society."[23]

Other challenges include making sure AI works in the real world, not just in test conditions. For example, an image-recognition model that works well in testing may fail if the lighting is poor or the camera is blurry when in actual use. AI systems also need to keep learning over time. A cybersecurity model that detects today's threats may be outdated tomorrow if it does not adapt. Another concern is energy use. Large-scale AI systems often require huge amounts of power, so engineers will need to design models that are more energy efficient in the future.

These challenges can seem daunting, but they also represent opportunities. If future AI engineers can figure out how to make systems fairer, more accurate, and greener, they will not just solve problems but also move the entire field forward. And for teens today, that means a chance to shape a future where AI improves lives, protects people, and helps society in ways we are only beginning to imagine.

SOURCE NOTES

Introduction: The New Frontier

1. Quoted Anna Tong and Kenrick Cai, "OpenAI, Google and xAI Battle for Superstar AI Talent, Shelling Out Millions," Reuters, May 21, 2025. www.reuters.com.
2. Quoted in Daniel Daines-Hutt, "How a ZTM Student Became a Machine Learning Engineer," *Zero To Mastery Blog*, May 15, 2024. https://zerotomastery.io.

Chapter One: What Does an AI Engineer Do?

3. Quoted in *Economic Times,* "Who Is Kairan Quazi, the 16-Year-Old Bengali Techie Who Left Elon Musk's SpaceX to Become a 'Wolf of Wall Street,'" August 21, 2025. https://economictimes.indiatimes.
4. Quoted in Meenakshi Sharma, "Not Just Animal Instinct, AI Now to Protect MP's Wildlife," *Times of India,* August 19, 2025. https://timesofindia.indiatimes.com.
5. Quoted in Ioana Patringenaru and Daniel Kane, "This Bespoke AI Tutor Helps Students Learn," *UC San Diego Today,* May 15, 2025. https://today.ucsd.edu.

Chapter Two: How Do You Become an AI Engineer?

6. Jason Crandall, "How I Became an AI Engineer," Medium, November 14, 2024. https://medium.com.
7. Quoted in El Dorado High School, "El Dorado Aztecs Tackle Real-World Problems at AI Hackathon," Socorro Independent School District, June 5, 2025. www.sisd.net.
8. Quoted in Holly Wojcik, "Penn Engineering Launches First Ivy League Undergraduate Degree in Artificial Intelligence," *Penn Today*, February 13, 2024. https://penntoday.upenn.edu.
9. Quoted in Canadian Broadcasting Corporation, "Why MIT Researcher Is Calling for 'Algorithmic Justice' Against AI Biases," May 12, 2025. www.cbc.ca.

Chapter Three: What Skills and Personal Qualities Matter Most—and Why?

10. Quoted in Byron Reese, "A Conversation with Sameer Maskey," *Voices in AI* (podcast), Episode 87, May 16, 2019. https://voicesinai.com.

11. Quoted in Jane Goodall, "Fascination, Persistence and Optimism: How Fei-Fei Li Helped Shape the AI Revolution in a Field Dominated by Alpha Males," The Conversation, February 5, 2024. https://theconversation.com.
12. Nadeen Matthews Blair, "Interested in a Career in AI? My Advice for Young People and Parents," LinkedIn, July 18, 2025. www.linkedin.com.

Chapter Four: What Is It like to Work as an AI Engineer?

13. Quoted in Agnes Applegate, "I'm a 28-Year-Old AI Engineer in Big Tech. Here's My Advice for Others Who Want to Break Into This Growing Field," *Business Insider,* July 10, 2025. www.businessinsider.com.
14. Quoted in Applegate, "I'm a 28-Year-Old AI Engineer in Big Tech."
15. Quoted in Hayden Field, "AI Engineers Report Burnout and Rushed Rollouts as 'Rat Race' to Stay Competitive Hits Tech Industry," CNBC, May 3, 2024. www.cnbc.com.

Chapter Five: Advancement and Other Job Opportunities

16. Quoted in Mark Samuels, "5 Ways Ambitious IT Pros Can Future-Proof Their Tech Careers in an Age of AI," Yahoo! Tech, October 25, 2025. https://tech.yahoo.com.
17. Quoted in OpenAI, "Vijaye Raji to Become CTO of Applications with Acquisition of Statsig," September 2, 2025. https://openai.com.
18. Quoted in Denys Krasnikov, "Grammarly Opens New Kyiv Office as Demand Rises for Help with English," *Kyiv Post,* July 6, 2018. www.kyivpost.com.

Chapter Six: What Does the Future Hold for AI Engineers?

19. Muhammad Tuhin, "9 Industries Being Revolutionized by AI," *Science News Today,* April 28, 2025. www.sciencenewstoday.org.
20. Quoted Helen Quinn, "The Transformative Use of AI in Radiotherapy and Beyond: Insights from Professor Raj Jena," Hospital Healthcare Europe, February 24, 2025. https://hospitalhealthcare.com.
21. Quoted in Imperial College Healthcare NHS Trust, "AI Could Predict Type 2 Diabetes Up to 10 Years in Advance," November 28, 2024. www.imperial.nhs.uk.

22. Quoted in David Elliott, "This AI Tutor Could Make Humans 10 Times Smarter, Its Creator Says," World Economic Forum, July 29, 2024. www.weforum.org.
23. Quoted in Stefan Milne, "AI Tools Show Biases in Ranking Job Applicants' Names According to Perceived Race and Gender," University of Washington, October 31, 2024. www.washington.edu.

Kriti Goyal works as a machine learning engineer in Seattle. This interview was conducted by email.

Q. Why did you become an AI professional?
A. My family is one of doctors, and the path of medicine was laid out for me; it was noble, respected, and understood. The world of technology, of Silicon Valley, might as well have been on another planet. The turning point wasn't a grand plan; it was a video my cousin showed me from Code.org. It had all these tech legends—Mark Zuckerberg and Bill Gates—but the message was what struck me. It said that with code, you don't need permissions, a factory or massive capital. Just a laptop and an internet connection to turn ideas into reality. I saw early on through school and my internships that AI was going to be the great disruptor, and I knew I didn't want to be a spectator; I wanted to be in the driver's seat.

Q. What does your typical workday look like?
A. My day splits into three parts. First, mornings are for disciplined curiosity—not just skimming papers but connecting dots between breakthroughs and real problems we're solving. This isn't always directly productive, but it's where innovation begins. The second part is collaboration mode—aligning with teams upstream, downstream (clients), my team, and management. We're whiteboarding, debating design, clarifying assumptions, and translating high-level vision into a written blueprint. This phase is all about building consensus and clarity. The third part, my favorite, is building. It's the state of deep work and flow, where the real creation happens—turning abstract math and architectural docs into functional reality, products that millions interact with. Honestly, that feeling is the reason I got into this field in the first place.

Q. What do you like most about your job?

A. The best part is knowing you are building at the absolute edge of tomorrow. There's a palpable feeling that the code you write today can fundamentally change how millions of people live their lives next year. I recently watched an interview with Fei-Fei Li, a pioneer in computer vision, who said her life's goal was to get an AI to tell a story from an image. That was once a distant dream; today, it's a reality. To witness and contribute to that pace of innovation is a privilege.

Q. What do you like least about your job?

A. The hardest part is the "creative destruction" inherent to that speed. In AI, you're building with state-of-the-art materials that are replaced every few months. It's like an architect designing a magnificent skyscraper, knowing that in a year, a new, stronger material will exist that makes their current design obsolete. You pour your heart into a project, and six months later, you might be the one dismantling it to make way for something even better. It teaches you to fall in love with the process of innovation itself, not just the final product—like building intricate sandcastles knowing the tide will come in.

Q. What personal qualities do you find most valuable for this type of AI work?

A. First principles thinking and the ability to stay with the problem and enjoy the challenge and thrill of solving something that is difficult; the ability to constantly learn and reinvent.

Q. What is the best way to prepare for this type of job?

A. I fought my way into a top engineering school and landed at Amazon working on a machine learning attribution model used to guide the apportionment of Amazon's $10 billion+ advertising budget into various marketing channels, such as social media, Google, email advertising. I then started as a software engineer at my current company and then switched to a machine learning role. Machine learning and AI are fields that require deep expertise based on the specific role. So, I chose the master's route. The playbook now might be different. The basics are the same: (1) be good at problem-solving, and (2) have some experience building AI products.

Q. What advice do you have for students who might be interested in this career?

A. The hard truth for graduates today is that the front door—the online application portal—is jammed. Sending four hundred résumés can feel like shouting into a black hole.

Benjamin Franklin said, "Either write something worth reading or do something worth writing." The "do something worth writing" option is the builder's path. Build in public. Don't just take a course—find an open-source project from a company you admire (Apple, Google, Meta, etc., all have them) and make meaningful contributions. The maintainers notice real contributions, and they're often the same people making hiring decisions. The "write something worth reading" option is the thought leader's path. Join the discourse. Write about problems you've actually solved. Give talks at conferences. Engage thoughtfully on X and LinkedIn. Yann LeCun, Andrej Karpathy, and other AI leaders actively respond to smart takes. One good, insightful thread can put you on their radar in a way a résumé never could.

AI4ALL
https://ai-4-all.org
AI4ALL is a nonprofit that works to increase diversity and inclusion in AI careers. The organization offers summer programs, mentors, internships, and online resources so that high school and college students can explore AI. Teens can also participate in virtual AI classes.

Association for the Advancement of Artificial Intelligence (AAAI)
www.aaai.org
The AAAI is a professional scientific organization that promotes research and advancement of AI. Its website provides information about conferences, publications about AI topics, and a job search engine. Teens can search and read about various AI news in the "AI Topics" section of the website.

Google AI
https://ai.google
Google AI provides Google's research and projects in AI. The website showcases its latest work in areas like machine learning, health, and climate. On this website, teens can read articles and learn how AI is being used in real-world applications. The site also provides links to free courses, such as its Introduction to Generative AI.

Khan Academy
www.khanacademy.org
Khan Academy offers free, interactive lessons in many subjects, including computer science, coding courses such as Python, and AI for education. Teens can take these courses at their own pace. The courses include lessons, quizzes, and videos.

MIT OpenCourseWare

https://ocw.mit.edu

MIT OpenCourseWare provides free access to college-level courses that include several focused on AI, such as Advances in Computer Vision and Mathematics of Machine Learning. The courses provide lecture notes, assignments, and videos by MIT professors. This can provide teens with insight into what college courses in these subjects would entail.

INDEX

Note: Boldface page numbers indicate illustrations.

PICTURE CREDITS

Cover: DC Studio/Shutterstock

6: Maury Aaseng
16: AndreyPopov/iStock
22: AndreyPopov/iStock
36: gremlin/iStock
42: Bevan Goldswain/iStock

Sources: AI Engineer at a Glance

- "AI Jobs on the Rise: Q1 2025 Labor Market Analysis," Veritone, May 15, 2025.
- "AI Engineer Salary," ZipRecruiter, 2025. www.ziprecruiter.com.
- "The Future of Jobs Report: 2023," World Economic Forum, 2023. www.weforum.org.

ABOUT THE AUTHOR

Leanne K. Currie-McGhee lives in Norfolk, Virginia, with her husband, Keith, daughters Grace and Sol, and dog, Delilah. She has been writing educational book for over two decades.